FLORA OF TROPICAL EAST AFRICA

PODOSTEMACEAE

H.J. Beentje

Aquatic herbs, looking like mosses, lichens or algae and usually growing on rocks or stones submerged in fast-flowing water. Plant base usually thalloid, variable in form, without vessels, sometimes without xylem, bearing buds on the margins and surface from which shoots arise; thallus attached to the substrate by haptera, flattened disc-like organs excreting a cement-like substance, and sometimes with filiform rootlets; stem simple or with abbreviated side-shoots, sometimes suppressed. Leaves absent or, where present, alternate, entire to dissected, linear, filiform, or scale-like, 3-ranked or 2-ranked, sometimes densely imbricate on branches and flowering shoots; exstipulate or sometimes with 2 tooth-like stipules. Flowers and fruits produced aerially at low water. Flowers bisexual, regular or irregular, small, solitary or in cymes, sometimes cleistogamous; bracteoles 2, enclosing the flower ('spathella') and then tearing irregularly at anthesis allowing the pedicel to elongate beyond the spathella, or spathella absent and bracts subtending the flower; tepals absent or 2–3(–5), ± connate or free, or consisting of a small annular scale; stamens hypogynous, 1–2(–4), with the filaments usually at least basally connate, dehiscing introrsely with a longitudinal slit. Ovary sessile or stalked, (1–)2(–3)-locular with as many ± basally connate (sub-)sessile stigmas; ovules many, anatropous, on thickened axile placentas. Fruit a stalked capsule, smooth or ribbed, dehiscing into 2–3 valves in dry air; seeds usually many, very small, often with mucilaginous testa.

42 to 50 genera with 275 species in the tropics and subtropics of especially America and Asia.

Included here are the *Tristichaceae* J.C.Willis, which Cusset & Cusset (see Adansonia 10, 2; 1988) uphold.

The following treatment owes much to the excellent work of Cusset & Cusset, who put African podostem taxonomy on a solid basis.

1. Leaves moss-like, 0.4–1 mm long, 0.5 mm wide; spathella absent; perianth of 3 equal or subequal tepals; stamen 1; stigmas 3; capsule opening by 3 valves . . . .   1. **Tristicha**

   Leaves filiform or scale-like, not moss-like; flower-bud enclosed in a spathella which ruptures at anthesis when pedicel lengthens; perianth of 2 minute tepals; stamens 1–2; stigmas 2(–4); capsule opening by 2 valves 2

2. Capsule smooth, dehiscing into 2 equal caducous valves; stamens 2 . . . .   4. **Leiothylax**

   Capsule ribbed, dehiscing into slightly unequal valves, one of which is caducous 3

3. Stamens 2, the filament branched from a common base . . . .   2. **Ledermanniella**

   Stamen 1, filament unbranched . . . .   3. **Sphaerothylax**

1

1. TRISTICHA

Thouars, Gen. Nov. Madagasc.: 3 (1806); C. Cusset & G. Cusset in Bull. Mus. Nat. Hist. Nat. B, Adansonia 10, 2: 169 (1988)

Annual or perennial herbs, freshwater aquatics, attached to rocks and stones in streams, rivers and near waterfalls; basal part thalloid, ribbon-like, branched, bearing short or long leafy shoots or abbreviated fertile shoots. Leaves small, spaced along sterile branches, dense and imbricate on flowering branches, arranged in 3 ranks, sessile, entire or divided. Flowering shoots subtended by 2 or more short leafy branches, surrounding a reduced stem terminated by the solitary flower. Flower terminal and/or axillary, subtended by 2 scarious bracts longer than the leaves. Perianth segments (tepals) 3, connate near base, scarious; stamen 1; ovary sessile, 3-locular, with 3 free filiform stigmas. Fruit a ribbed capsule, septicidally 3-valved; seeds many, small.

Tristicha trifaria (*Willd.*) *Spreng.*, Syst. Veg. ed. 16, 1: 22 (1824); Baker & C.H. Wright in F.T.A. 6, 1: 121 (1909); G. Taylor in F.W.T.A. ed. 2, 1: 124, t. 43 (1954); Obermeyer in F.S.A. 13: 206, t. 30/1 (1970); C. Cusset in F.Z. 9.2: 8, t. 4 (1997); M.G. Gilbert in Fl. Eth. 2, 1: 191 (2000). Type: Mauritius, *Bory de St Vincent* s.n. (B-W, holo., BM!, P!, iso.)

Aquatic perennial herb; thalloid part ribbon-like, 1–4 mm wide. Long stems to 30 cm long, freely branched, producing leafy branchlets and short flowering shoots; leaves 3-ranked, sessile, elliptic to slightly obovate, 0.4–3.2 mm long, 0.5–0.6 mm wide, apex obtuse to rounded, main nerve ± visible. Flowers terminal and/or axillary, each subtended by 2 free leafy bracts; bracts longer and thinner than leaves; pedicel erect, 1–15 mm long, to 20 mm long in fruit. Tepals 3, scarious, connate for about half their length, free lobes 1–2 mm long, 0.4–0.7 mm wide; stamen 1(–2), filament 1.5–3 mm long, anther oblong, 0.5–1 mm long; ovary sessile, ovoid to ellipsoid, 0.5–1 mm long, 0.6–0.8 mm in diameter, with 3 filiform stigmas 0.2–0.7 mm long at apex. Fruit a 9-ribbed capsule 1.5–2.2 mm long, to 1.2 mm in diameter, dehiscing into 3 equal caducous valves; seeds many, small, orange-brown; testa with anastomosing ribs. Fig. 1 (page 3).

UGANDA. West Nile District: Koboko, June 1936, *A.S. Thomas* 1992!; Bunyoro District: Nkusi River, Sep. 1977, *Lye & Katende* 22864!; Mbale District: Elgon, Masaba, Dec. 1936, *A.S. Thomas* 2103!

KENYA. Fort Hall/Machakos District: Fourteen Falls, Jan. 1960, *Verdcourt* 2604!; N Kavirondo District: Malikisi R. 23 km ESE of Tororo, Oct. 1959, *G. Taylor* s.n.!; Masai District: Ol Doinyo Orok, *Napper* 854!

TANZANIA. Kigoma District: Kasye Forest, Mar. 1994, *Bidgood et al.* 2818!; Kilosa District: Great Ruaha R., 3 km S of junction with Yovi R., Sep. 1970, *Thulin & Mhoro* 889!; Iringa District: Ruaha R., Mayoge, Oct. 1970, *Richards & Arasululu* 26281!

DISTR. **U** 1–4; **K** 3–6; **T** 1, 2, 4, 6, 7; from West Africa to Ethiopia and south to South Africa; also in Madagascar, Mauritius, India, Australia, and the Americas

HAB. On rocks in streams and rivers and at top or base of falls, usually submerged but sometimes only partly so; pedicel elongating when plant no longer submerged; mat-forming, may be locally common; 450–2200(–2650) m

CONSERVATION NOTES. Least concern (LC)

SYN. *Dufourea trifaria* Willd., Sp. Pl. 5: 55 (1810)
     D. alternifolia Willd. in Ges. Naturf. Freunde Berlin Mag. N. Entdeck. Gesamten Naturk. 6: 64 (1814). Type: Madagascar, East, *Du Petit Thouars* s.n. (B-W, holo., BM!, P!, iso.)
     Tristicha alternifolia (Willd.) Spreng., Syst. Veg. ed. 16, 1: 22 (1824); Bak. & Wright in F.T.A. 6, 1: 121 (1909); Hauman in F.C.B. 1: 220 (1948)
     T. hypnoides Spreng., Syst. 4, Cur. Post.: 10 (1827); Bak. & Wright in F.T.A. 6, 1: 121 (1909). Type: Brazil, *St. Hilaire* s.n. (P, holo.)

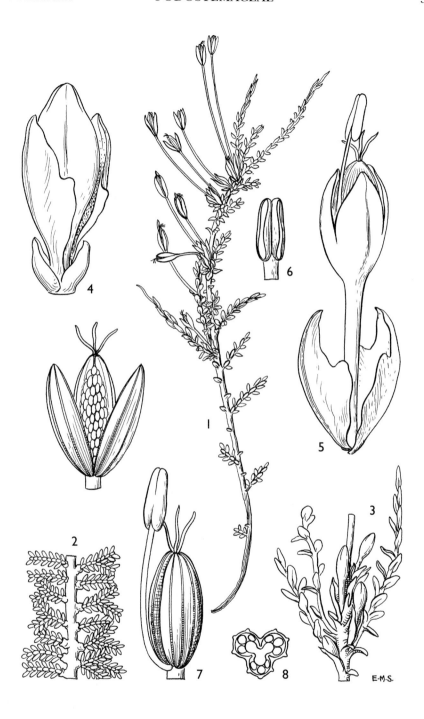

Fig. 1. *TRISTICHA TRIFARIA* — **1**, habit, × 3; **2**, habit showing leaves, × 3; **3**, leaves and unopened flowers, × 6; **4**, unopened flower, × 24; **5**, mature flower, × 18; **6**, anther, × 24; **7**, flower without tepals, × 24; **8**, cross-section of fruit, diagrammatic; **9**, split fruit, × 24. 1, 3–9 from *Adames & Akpabla* 4287, 2 from *Ross* 303. Drawn by Margaret Stones, from the Flora of West Tropical Africa.

T. alternifolia (Willd.) Spreng. forma *sambesiaca* Engl., E.J. 60: 456 (1926). Type: Zimbabwe, Victoria Falls, Livingstone Is., *Engler* 2947b (B, holo.) (type is mixed with *Ledermanniella tenax*)

NOTE. The flowering period is very brief – it takes about 3 days from flower bud to ripe fruit.

Within a population there may be several forms, the lowermost plants have elongated branches, while the higher-up plants are more compact or even just crustose.

All our material is subsp. *trifaria*; Cusset & Cusset have a subsp. *pulchella* (Weddell) C.Cusset & G.Cusset from West Africa.

2. LEDERMANNIELLA

Engl. in E.J. 43: 378 (1909); C. Cusset in Bull. Mus. Nat. Hist. Nat. B, Adansonia 5, 4: 361–390 (1983) and 6, 3: 249–278 (1984)

Inversodicraeia R.E.Fr., Wiss. Ergebn. Schwed. Rhod.–Kongo Exped. 1, 1: 56 (1914), as *Inversodicraea*
Monandriella Engl. in E.J. 60: 457 (1926)

Perennial herbs, submerged freshwater aquatics; basal part thalloid, leafy, ribbon-like or deeply divided; shoots with well-developed usually branched stems. Leaves exstipulate, long and linear and then usually divided, or reduced and scale-like, borne on stems and branches or on fertile shoots only. Flowers terminal and solitary or occasionally also axillary to upper branches, reflexed in bud within the protective bracts (spathella); spathella tearing irregularly at anthesis, the elongating pedicel exserting, bearing the then erect flower. Tepals 2, linear or filiform; stamens 1–2(–3), the filaments connate in their lower part; ovary sessile or stalked, 1-locular with central placentation; stigmas 2, usually free. Capsule usually 8-ribbed, splitting into 2 equal carpels; seeds many, small, dorsiventrally flattened, with reticulate testa.

African genus with 44 species.

1. Plants bearing scale leaves, at least near apex of branches   1. *L. tenax*
   Plants without scale leaves . . . .   2. *L. maturiniana*

Note: *L. ramossissima* C.Cusset is known from a single collection just over the border in the Sudan (Imatong Mts: *A.S. Thomas* 1660); the thallus is unknown, the leaves are only known from 2–3 mm long linear leaves on the terminal branches. This would key out as *L. maturiniana* but differs in the undivided leaves and the larger tepals (± 0.6 mm long).

1. **Ledermanniella tenax** (*C.H.Wright*) *C.Cusset* in Adansonia ser. 2, 14(2): 275 (1974) & in Bull. Mus. Nat. Hist. Nat. B, Adansonia 5, 4: 384 (1983) & in F.Z. 9, 2: 4, t. 1b (1997). Type: Zambia, Victoria Falls, Livingstone Is., *Kolbe* 3149 (K!, holo., BM!, BOL, iso.)

Perennial herb, much branched; thalloid part ribbon-like, ± 1 mm wide; stem simple or branched, slightly fleshy, to 20 cm long. Scale-like leaves scattered on stem, densely imbricate on fertile branches, 3-dentate, 0.8–1.5 mm long, 0.3–0.5 mm wide, the teeth 0.2–0.7 mm long; scale leaves just below spathella longer and narrower than those on stem. Filiform leaves dichotomously or trichotomously divided into filiform segments, 3–5 cm long (not seen in East African specimen). Bracts and bracteoles 1.5–2 mm long, dichotomously branched with filiform segments. Flowers 1–several on very short uppermost branches, and 1–2 axillary to branches; spathellas ellipsoid to obovoid; pedicel elongating to 8 mm after anthesis. Tepals 2, filiform, 0.5 mm long; stamens with filaments 1.5 mm long, connate for half their length, anthers 1.5 mm long; ovary ellipsoid, 2 mm long, 1.2 mm in diameter. Capsule ellipsoid, 1.7–2.5 mm long, 8-ribbed; seeds ellipsoid, 0.4 mm long, 0.16 mm in diameter.

FIG. 2. *LEDERMANNIELLA MATURINIANA* — **1**, habit, × ²⁄₃; **2**, habit detail with thalloid base, × 2; **3**, leaves, × 6; **4**, emerging flower, × 16; **5**, emerging flower, × 10; **6**, flower at anthesis, × 10; **7**, mature flower, × 12; **8**, seeds on central capsule stalk, × 16. 1 & 8 from *Napper* 409, 2 & 4 & 6 from *Gillett* 19270, 3 & 5 & 7 from *Verdcourt* 719. Drawn by Juliet Williamson.

TANZANIA. Iringa District: Lupembe, Ruhudje R., Aug. 1931, *Schlieben* 1131a!
DISTR. **T** 7; Angola, Zambia, Zimbabwe, Botswana, Namibia
HAB. Mat-forming on rocks in fast-flowing water, near water-level with the flowering stems emerging; ± 1700 m
CONSERVATION NOTES. Least concern (LC)

SYN. *Dicraeia tenax* C.H.Wright in F.T.A. 6, 1: 125 (1909), as *Dicraea*
    Inversodicraeia tenax (C.H.Wright) R.E.Fr., Wiss. Ergebn. Schwed. Rhod.–Kongo Exped. 1,
    1: 56, t. 11, fig. 15–21 (1914); Engler, V.E. 9, 3, 1: 274 (1915) & in E.J. 60: 463 (1926),
    as *Inversodicraea*

2. **Ledermanniella maturiniana** *Beentje* **sp. nov.** species habitu cum *L. bifurcata* congruens sed foliis maioribus estipulatis atque e basi non prope apicem divisis, stigmatibus clavatis non filiformibus differt. Type: Kenya, Kiambu/Machakos District: Fourteen Falls, *Verdcourt* 719 (K!, holo., EA, iso.)

Perennial herb, branched; thalloid part foliaceous, to at least 7 mm wide; stem simple or branched, slightly fleshy, to 11 cm long. Scale leaves absent; filiform leaves ± distichous, branched 3–4 times from near base, 0.9–6.5 cm long, the segments 0.1–0.3 mm wide, flattened, the base half-sheathing. Bracts and bracteoles ?similar to leaves. Flowers 1–many on upper branches, terminal and axillary; spathella ellipsoid, slightly flattened, to 3.2 × 1.9 mm, tearing irregularly; pedicel elongating to 3–9 mm at anthesis. Tepals 2, subulate, 0.1–0.5 mm long; stamens 2, the filaments connate for 0.5–1.5 mm, free for 0.3–0.7 mm, anthers 0.7–0.8 mm long, 0.3–0.5 mm wide; ovary ellipsoid, 1.2–1.7 mm long, 0.8–1.1 mm in diameter; gynophore 0–0.25 mm long; stigmas club-shaped, 0.2–0.6 mm long. Capsule ellipsoid, 1.3–2 mm long, 8-ribbed; seeds ellipsoid, 0.1–0.2 mm long. Fig. 2 (page 5).

KENYA. Kiambu/Machakos District: Fourteen Falls, Oct. 1955, *Napper* 409!; Kitui District: Kindaruma Dam lake, W end, Dec. 1970, *Gillett* 19270!; Trans-Nzoia District: Mt Elgon, Suam R at Karamoja Drift, Feb. 1935, *G.Taylor* 3140!
DISTR. **K** 3–5; not known elsewhere
HAB. On slightly submerged rocks in waterfall or fast-flowing stream, flowering when rocks are ± exposed, in full sunlight, locally abundant; 800–1500 m
CONSERVATION NOTES. As this occurs in a limited distribution area, and pollution in waterways from where it has been collected is now a serious issue, I would suggest this as endangered, EN (B1a, b)

NOTE. This species resembles *Ledermanniella bifurcata* (Engl.) C.Cusset from Cameroon but differs in the leaves dividing from near base and not only near the apex, the leaves being 0.9–6.5 cm long, not 0.7–2.5 cm, and in lacking visible stipules; in the stigmas being club-shaped and squat and 0.2–0.6 mm long, not filiform and 0.6–0.7 mm.
    The specific epithet honours Dr S. Maturin, physician and natural philosopher.
    Specimens from Western Kenya have longer filaments (connate for 1.2–1.5 mm rather than the Eastern 0.5–0.7 mm) and slightly longer stigmas than those from the E part of the country.

3. **SPHAEROTHYLAX**

Bisch. in Flora 27: 426, t. 1 (1844)

Anastrophaea Wedd. in DC., Prodr. 17: 78 (1873)

Perennial herbs, submerged freshwater aquatics; thalloid base attached to substrate, foliaceous and lobed or ribbon-like and branched; stems absent or very short, to elongate and branched with leaves and flowers congested at the nodes. Leaves dissected into filiform segments. Spathellas ± sessile on the thalloid part or in clusters axillary to leaves on elongated stems, solitary or aggregated, subtended by 2 scale-like bracts, surrounding the reflexed flower bud until anthesis, when the flower breaks the spathella and the pedicel becomes exserted, with an erect flower. Flower with 2 tepals, these minute and subulate; stamen 1; ovary 1-locular, subspherical, with 2 subulate stigmas. Capsule subspherical, 8-ribbed, dehiscing into 2 equal or slightly unequal valves, the smaller one caducous; seeds many, black, flattened and ovoid.

2 species in Africa.

Note. The number of stamens has been debated for this genus. The protologue says that there are three, but then Bischoff counted the two tepals as staminodes; otherwise, from early days they have been counted as two, with their filaments fused to the apex. Taylor (in J.B. 76: 111, 1938) says the stamen structure consists of a single stamen with a broad filament and anther. The filament is so expanded towards the apex that the lobes of the deeply cleft anther appear separate. Cusset follows Taylor, as I do here.

The author of the genus has been cited as Krauss in many publications, but it is definitely Bischoff.

1. Leaf segments 0.3–1.3 mm wide; anthers 0.5–0.7 mm long . . .   1. *S. abyssinica*
   Leaf segments 0.1–0.3 mm wide; anthers 0.3–0.35 mm long . .   2. *S. algiformis*
   (see note under *S. algiformis* – these taxa are close)

1. **Sphaerothylax abyssinica** (*Wedd.*) *Warm.*, Fam. Podost. iv: 14, 39, fig. 17 (1891); Oliv. in Hook. Icon. Pl. 24: t. 2356 (1895); Baker & C.H. Wright in F.T.A. 6, 1: 128 (1909); M.G. Gilbert in Fl. Eth. 2, 1: 194, t. 26.2 (2000). Type: Ethiopia, Begemder, Gafat, *Schimper* 1181 (B, holo., BM!, K!, iso.)

Herb with thalloid base, irregularly lobed, to 4.5 cm wide, dark translucent olive green or blood-red; stem ± unbranched to much branched, green, to 30 cm long, at intervals bearing leaves or contracted side-branches which bear leaves and clusters of flowers. Leaves elongate, 2–3-divided, narrowly linear, 2–15 cm long, segments 0.3–1.3 mm wide, sometimes deciduous at time of flowering. Flowers both solitary on thallus and in glomerules on the contracted side-branches, or in up to 10 mm long rows of groups of flowers along the main stem; spathella sessile, ellipsoid, 2.1–3 mm long, 1.3–1.9 mm in diameter, subtended by subulate bracts to 0.8 mm long; pedicel pink to red, lengthening to 1.5–5 mm, slightly winged; tepals at base of ovary, subulate, 0.2–0.5 mm long; stamen with filament 0.5–1.7 mm long, anthers 2, 2-locular, pale green or greenish yellow, 0.5–0.9 mm long, one set slightly higher than the others and slightly diverging; ovary brown-red or red, obovoid or ellipsoid, 1–1.5 mm long, 0.7–1.2 mm in diameter, sessile or stalked for up to 0.2 mm; stigmas red, filiform, 0.3–0.6 mm long. Capsule ellipsoid, 1.2 mm long, 0.9 mm in diameter, 8-ribbed. Fig. 3 (page 8).

KENYA. Embu District: Thuchi R. by old main road, Dec. 1975, *Dransfield* 4814!
TANZANIA. Iringa District: Kitulo Plateau, above Salala waterfalls, May 1987, *Iversen et al.* 87/706!; Njombe District: Ruhudji R. waterfall, July 1956, *Milne-Redhead & Taylor* 11028!; Rungwe District: Kiwira R., June 1969, *Wingfield* 292!
DISTR. **K** 4; **T** 7; Ethiopia
HAB. On rocks in fast-flowing streams or in rocks close by waterfalls; 1300–2400 m
CONSERVATION NOTES. Probably least concern (LC)

SYN. *Anastrophea abyssinica* Wedd. in DC., Prod. 17: 79 (1873); Engl. in E.J. 60: 466 (1926)
   Dicraeia violascens Engl. in E.J. 30: 312 (1902). Type: Tanzania, Njombe District, Ukinga Mts, Liroro [Diroro] R., *Goetze* 943 (B†, holo., BM!, iso.)
   Sphaerothylax sanguinea Chiov. in Ann. Bot. 1911: 131 (1911). Type: Ethiopia, Debarek, *Chiovenda* 2939 (FT, holo., BM!, iso.)
   Leiocarpodicraeia violascens (Engl.) Engl. in E.J. 38: 98 (1905) & 60: 464 (1926)
   Leiothylax violascens (Engl.) C.H.Wright in F.T.A. 6, 1: 125 (1909)

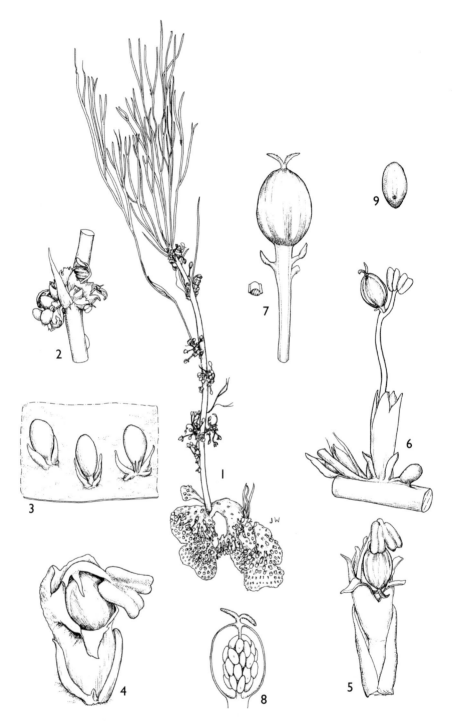

Fɪɢ. 3. *SPHAEROTHYLAX ABYSSINICA* — **1**, habit with thalloid base, × ²/₃; **2**, flower cluster along stem, × 4; **3**, solitary flowers on thallus, × 6; **4**, emerging flower, × 16; **5**, emerging flower, × 12; **6**, mature flower, × 8; **7**, ovary with tepals, × 16; **8**, opened fruit, × 16; **9**, seed, × 32. 1, 3–9 from *Ash* 2612, 2 from *Iversen et al.* 87/706. Drawn by Juliet Williamson.

NOTE. C. Cusset in Adansonia sér. 2, 20, 2: 199–207 (1980), in her thorough treatment of *Leiothylax*, said it was difficult to gauge the taxonomic value of *L. violascens* as 'all the material cited in the protologue has disappeared' – but a duplicate is present at BM, and is a close match for *Milne-Redhead & Taylor* 11028. Cusset thought the 'Kinga Mts, Divoro' locality was in Malawi, but did not mention the taxon in her treatment in FZ.

2. **Sphaerothylax algiformis** *Bisch.* in Flora 27: 426 (1844); R.E. Fr., Wiss. Ergebn. Schwed. Rhod.–Kongo Exped. 1, 1: 56, t. 11, fig. 1–14 (1914); Obermeyer in F.S.A. 13: 209, t. 31/2 (1970); C. Cusset in F.Z. 9, 2: 6, t. 3 (1997). Type: South Africa, near Pietermaritzburg, *Krauss* s.n. (B†, holo.)

Herbs with thalloid creeping body resembling *Marchantia* and foliaceous or ribbon-like, green or dark red, tightly adhering to the substrate; stems erect, woody, simple or branching, reddish, 5–50 cm long with side-branches contracted at intervals, these side-branches bearing the leaves and clusters of flowers. Leaves filiform, branching dichotomously, 2–15 cm long, the segments 0.1–0.3 mm wide. Flowers solitary and borne on the thalloid body or in clusters axillary to leaves, subtended by 2–3 unequal bracts, these ovate and convex; spathella spherical to obovoid, splitting irregularly at anthesis; flower recurved within spathella, becoming erect at anthesis, pedicel red, slightly winged, lengthening to 1–4 mm; perianth of 2 minute subulate tepals on each side of the androphore; stamen 1, filament 0.5–1 mm long, one anther slightly higher than the other, 0.3–0.35 mm long; ovary ovoid, sessile; ovary ellipsoid, 0.9–1 mm long, 0.7–0.8 mm in diameter; stigmas ovate and rather thick, 0.1–0.2 mm long. Capsule ovoid, 0.8–1.25 mm long, 0.75–1 mm in diameter, 8–12-ribbed, splitting into 2 equal or slightly unequal valves, one caducous; seeds many, slightly flattened, black, reticulate.

TANZANIA. Mpanda District: Mpanda–Uvinza road, 05°51' S, 30°42' E, June 2000, *Bidgood et al.* 4710!
DISTR. **T** 4; Zambia, Malawi, Zimbabwe, South Africa
HAB. On rocks in fast-flowing stream; 1275 m
CONSERVATION NOTES. Least concern (LC)

SYN. ?*S. wageri* G.Taylor in J.B. 76: 112 (1938). Type: South Africa, Lydenburg District, Sabie, *Wager* s.n. (BM!, holo., PRE, iso.)

NOTE. Taylor (in his description of *S. wageri*) cites *Greenway* 3515 from Njombe, but this is more likely *S. abyssinica*. The type of *S. wageri* has few leaves, though they conform to the description of *S. algiformis*, but the anthers are up to 0.5 mm long, and so resemble those of *S. abyssinica*!

4. LEIOTHYLAX

Warm. in Kongel. Danske Vidensk. Selsk. Skrift., Naturvidensk. Math. Afd., ser. 6, 9 (2): 147 (1899); C. Cusset in Adansonia sér. 2, 20, 2: 199–207 (1980)

Dicraeia sect. *Leiocarpodicraeia* Engl. in E.J. 20: 234 (1894)
Leiocarpodicraeia (Engl.) Engl. in E.J. 38: 94, 98 (1907)

Perennial herbs, submerged freshwater aquatics; thalloid base attached to substrate, foliaceous and lobed or ribbon-like and branched; stems elongate and branched. Leaves dissected into linear segments. Spathellas in clusters along the stems, solitary or aggregated, subtended by 2 scale-like bracts surrounding the reflexed flower bud until anthesis, when the flower breaks the spathella and the pedicel becomes exserted, with an erect flower. Flower with 2 tepals, these minute and subulate; stamens 2; ovary 1-locular, subspherical, with 2 subulate stigmas. Capsule globose, smooth, dehiscing into 2 equal caducous valves; seeds many, black, flattened and ovoid.

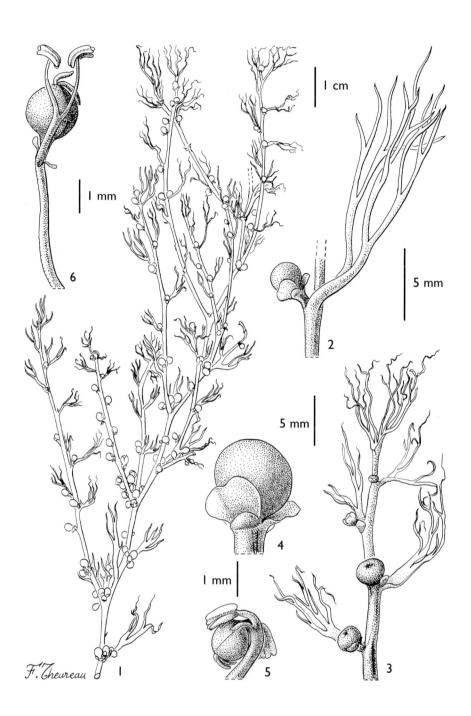

Fig. 4. *LEIOTHYLAX DRUMMONDII* — **1**, habit; **2**, flower bud on leafy shoot; **3**, branched shoot apex; **4**, unopened flower; **5**, unopened flower, spathella removed; **6**, flower at anthesis. Drawn by F. Theureau, from Flora Zambesiaca.

3 species in Africa.

Leiothylax drummondii *C.Cusset* in Adansonia sér. 2, 20, 2: 204, t. 1 (1980) & in F.Z. 9, 2: 6, t. 2 (1997). Type: Zambia, Mubalashi R., Kapiri Mposhi–Mkushi road, *Drummond* 8271 (P!, holo., SRGH, iso.)

Thalloid part foliaceous, lobed; stems caespitose, to 50 cm long, branched. Leaves dividing dichotomously 2–3 times, 2–6.5 cm long, the leaf segments linear, 0.1–?2 mm wide. Spathellas in clusters along the stem, subspherical, sessile, 1–2.5 mm in diameter; pedicel elongating to 7–9 mm at anthesis and to 10 mm in fruit. Tepals 2, minute, subulate, 0.3–0.4 mm long; stamens 2, the filaments up to 0.8 mm long, joined for over half their length; ovary globose, 1.1–1.2 mm in diameter, on a stalk 0.3–1.5 mm long; stigmas 2–4, linear, 0.7–0.8 mm long. Capsule globose, smooth, dehiscing by 2 equal caducous valves; seeds many, black, ± 0.25 mm long and 0.1 mm wide. Fig. 4 (page 10).

TANZANIA. Iringa District: Ruaha National Park, Mwayangi, Great Ruaha R., Aug. 1969, *Greenway & Kanuri* 13774!
DISTR. **T** 7; central Zambia
HAB. On rocks in fast-flowing river; ± 750 m
CONSERVATION NOTES. Only known from 2 localities, over a 1000 km apart; data deficient (DD)

NOTE. Part of a mixed collection; the K sheet was among the indeterminates and also had some *Tristicha trifaria* on it.

The Tanzanian material has branched stigmas, so appearing as four rather than the two Cusset describes.

INDEX TO PODOSTEMACEAE

New names validated in this part

Ledermanniella maturiniana *Beentje*

PLANTS PEOPLE
POSSIBILITIES

First published in 2005 by
Royal Botanic Gardens, Kew
Richmond, Surrey, TW9 3AB, UK
www.kew.org

ISBN 1 84246 110 9

Design by Media Resources, typesetting and page layout by Margaret Newman,
Information Services Department,
Royal Botanic Gardens, Kew.

Printed by Cromwell Press Ltd.

For information or to purchase all Kew titles please visit
www.kewbooks.com or email publishing@kew.org